From Echoes to Insights

Radar Technology's Unseen Influence on Scientific Discovery and Technological Advancements

Leola Carter

Contents

1 Introduction to Radar

Early radar development was a fascinating blend of scientific ingenuity and wartime necessity. The initial experiments, often conducted in relative secrecy, involved rudimentary technology compared to today's sophisticated systems. Scientists like Robert Watson-Watt, credited with the invention of practical radar, faced significant challenges in overcoming limitations in signal processing and generating sufficiently powerful radio waves to achieve effective detection. Their initial successes, even with limited range and accuracy, proved invaluable. This early work laid the foundation for the exponential growth that would transform radar from a niche scientific pursuit to an essential technology for military and civilian applications. The transition from rudimentary vacuum tubes to more efficient components significantly improved the performance and portability of radar systems.

Understanding how radar works requires grasping its core principles. A radar system transmits electromagnetic pulses, which then bounce off targets like aircraft or raindrops. The reflected signals, or echoes, are received by the radar system. By precisely measuring the time it takes for the pulse to travel to the target and return, the distance to the target can be calculated.

The strength of the returned signal indicates the target's size and reflectivity. This is fundamentally similar to sonar, but radar uses radio waves, allowing for much greater range and versatility due to the radio waves' ability to travel through the atmosphere without significant attenuation, unlike sound waves. The angle of the received signal relative to the radar antenna's position helps determine the target's bearing. The combination of distance and bearing provides a precise location of the target, generating the data visualized on the radar screen. This simple yet powerful methodology has been refined and adapted for a vast array of applications.

The Second World War dramatically accelerated the development and deployment of radar. Initially used by the British in their defense against German air raids, it rapidly became a critical tool across all branches of the military. Airborne radar allowed for the early detection of enemy aircraft, enabling timely responses and significantly improving the success rate of interceptor missions. Naval radar enhanced ship-to-ship combat, enabling more accurate targeting and improved navigation in challenging conditions. Ground-based radar systems provided early warning of incoming attacks, giving vital time to prepare defenses and mobilize troops. The effectiveness of radar in World War II highlighted its strategic importance, propelling further research and technological advancements in the post-war era. Its impact on the war's outcome is undeniable, saving countless lives.

Following the conflict, several crucial advancements occurred. Miniaturization became a major focus, leading to smaller and more portable radar systems suitable for a wider range of applications. The development of more powerful and efficient components, such as transistors and integrated circuits, was key to this process. These improvements significantly reduced the size, weight, and power consumption of radar systems, expanding their potential use beyond military applications. Simultaneously, accuracy and reliability improved dramatically, enabling more precise measurements and reducing false alarms. This surge of innovation transformed radar into a versatile tool with applications ranging from air traffic control to weather forecasting. This transition from large, power-hungry military systems to smaller, more efficient civilian systems significantly expanded the use of radar technologies.

The integration of digital signal processing has been revolutionary. Modern radar systems often use advanced algorithms to filter out noise, improve target detection, and even identify different types of targets based on their radar signature. Digital signal processing allows for sophisticated signal analysis, enabling finer discrimination between targets and the background environment. This leads to enhanced detection and tracking capabilities, particularly in cluttered environments. For example, in air traffic control, this allows for the safe and efficient management of many aircraft in close proximity. Similarly, the ability to filter clutter, such as ground echoes or weather formations, is critical for efficient detection and analysis in several applications. Furthermore, modern radar systems now frequently use phased array antennas, which enable electronic beam steering without physically moving the antenna, providing greater flexibility and speed in tracking multiple targets simultaneously.

1.1 Early Development

The earliest explorations into what we now know as radar weren't driven by a singular, focused goal. Instead, it was a confluence of scientific curiosity and the pressing need for improved naval navigation that laid the groundwork. Early experiments, often conducted independently by researchers in various countries, focused on the detection of objects using radio waves. These weren't sophisticated systems; think crude antennas and rudimentary receivers, painstakingly piecing together echoes to determine a target's distance and direction. The challenges were immense: filtering out noise, ensuring reliable signal transmission over long ranges, and accurately interpreting the often-faint reflections. These were not easily overcome.

Christian Hülsmeyer's 1904 patent for a "telemobiloscope" stands as a pivotal early contribution. This device, while rudimentary by modern standards, demonstrated the potential of radio waves to detect the presence of ships in foggy conditions – a significant breakthrough for maritime safety. His invention, however, did not achieve widespread adoption. The technology simply wasn't mature enough; the limitations in generating and receiving high-frequency signals posed significant practical barriers. The world was not yet ready for such a complex system.

World War I spurred further research, but progress remained incremental. The need for improved detection systems was evident, yet the technological hurdles were considerable. The war highlighted the critical role of detection in warfare, creating a catalyst for increased investment and innovation. But it wasn't just about detecting enemy combatants; research also explored the potential of radio waves for ranging – determining the exact distance to a target. This aspect proved to be significantly more challenging, requiring a much deeper understanding of radio wave propagation and signal processing.

The interwar period saw advancements in radio technology – improvements in vacuum tube technology and signal processing circuits – providing the foundation for more capable radar systems. Researchers began to grasp the critical role of pulse modulation, where short bursts of radio waves were transmitted and their reflections analyzed. This technique improved the accuracy of distance measurements and dramatically reduced interference. The theoretical groundwork was being laid, gradually paving the way for the revolutionary impact of radar during the following conflict.

The breakthroughs weren't solely technological. Theoretical advancements in signal processing techniques were equally vital. Mathematicians and physicists were essential in the development of the algorithms and methods necessary to extract meaningful information from the noisy radar signals. Early signal processing techniques were limited by the available computing power, which emphasized the development of efficient and robust algorithms that could be implemented using the available electronics. The focus shifted from simple detection to extracting information like speed and size. The challenge wasn't merely to detect the echo – it was to understand what the echo meant.

The development of the magnetron, a vacuum tube capable of generating high-power microwaves, proved to be a game-changer. This invention, along with advancements in receiver sensitivity, dramatically increased the range and accuracy of radar systems. The magnetron allowed for the creation of radar systems capable of detecting aircraft at significant distances,

leading to significant improvements in air defense capabilities. The implications were profound, and it would quickly revolutionize aerial warfare. Suddenly, the previously invisible became detectable, changing tactics and strategies overnight.
The early development of radar was a collaborative effort. Researchers from different fields — physicists, engineers, mathematicians — contributed their expertise, pushing the boundaries of what was thought possible. It was a long and complex journey, one that involved numerous false starts, dead ends, and unexpected breakthroughs. The transition from crude experiments to sophisticated systems was gradual, but the cumulative effect was transformational, altering the course of both war and peace. The story of early radar is a testament to the power of human ingenuity and the profound impact of seemingly esoteric scientific discoveries. The early years weren't just about creating a machine; they were about creating a new way to see the world.

1.2 Basic Principles

Electromagnetic waves form the bedrock of radar technology. These waves, invisible to the human eye, are emitted by a transmitter and travel outward at the speed of light. When they encounter an object, a portion of the energy reflects back towards the source. This reflected signal, or echo, carries vital information about the object's size, shape, speed, and distance—all critical elements in deciphering what the radar is "seeing". Understanding this fundamental principle is crucial to grasping the broader capabilities of radar.
The time it takes for the transmitted wave to return—the round-

trip time—is directly proportional to the distance of the target. Simple trigonometry, utilizing the speed of light as a constant, allows us to calculate this distance precisely. This basic principle, though seemingly straightforward, underpins the entire operation of most radar systems. More complex systems utilize additional methodologies, yet this remains the foundational concept.

Moreover, the strength of the returning signal, or echo intensity, reveals information regarding the size and reflectivity of the target. A larger object, or one with a more reflective surface, will generally produce a stronger echo. This simple relationship forms the basis for distinguishing between various targets and determining their relative sizes. Understanding this property also illuminates the limitations of radar systems, such as challenges posed by camouflage or small, low-reflectivity objects.

Furthermore, the Doppler effect plays a crucial role in radar. This effect, which causes a change in the observed frequency of a wave due to relative motion between the source and the observer, allows radar systems to measure the radial velocity of a target. This is crucial for applications like weather tracking and traffic monitoring, where the speed of objects is a primary concern. The shift in frequency, either higher or lower, directly correlates with the target's movement towards or away from the radar.

Signal processing techniques are essential for extracting meaningful information from the received signals. Raw radar signals are often obscured by noise and interference. Various signal processing algorithms are used to filter out this noise and enhance the desired echoes. These algorithms, ranging from simple filters to complex digital signal processing techniques, are pivotal in improving the accuracy and reliability of radar systems, making weak signals easier to process and improving the resolution in a cluttered environment. Mastering these techniques is essential for any serious endeavor in radar technology.

Analyzing radar data requires a careful interpretation of the echoes and their characteristics. Factors such as signal-to-noise ratio, echo intensity, and Doppler shift are carefully considered to determine the properties of the target. This analysis often involves sophisticated mathematical models and computer algorithms. This careful analysis is crucial for transforming raw data into useful information, from the precise position of an aircraft to the intensity and direction of a weather system. The quality of data analysis directly dictates the reliability and value of radar applications.

The design of a radar system involves a careful selection of parameters, such as frequency, pulse width, and antenna characteristics. Each parameter significantly impacts the performance of the system. These parameters influence the range, resolution, and accuracy, and their selection is influenced by factors like the intended applications. A system designed for detecting long-range aircraft will have different parameters compared to one used for short-range obstacle detection. Understanding and controlling these parameters is paramount in designing effective radar systems.

Finally, radar technology is not a static field. Continuous advancements are being made in various aspects of radar systems. These advancements include improved signal processing techniques, more sophisticated antenna designs, and the incorporation of new technologies, leading to enhanced performance, range, and versatility. This ongoing evolution means that a comprehensive understanding of the fundamental principles will enable professionals to adapt to technological changes and leverage new advances. Staying informed about cutting-edge developments is crucial for future success in this dynamic field.

2 Radar in World War II

The Second World War witnessed a dramatic shift in warfare, a revolution driven in part by the unprecedented deployment of radar technology. Initially, rudimentary systems offered a limited but invaluable edge, allowing early detection of enemy aircraft, a critical advantage for defense. This early warning system, however crude, proved its worth countless times, providing precious minutes, sometimes even hours, to prepare for incoming attacks. These were vital moments which often meant the difference between survival and devastation. The impact was profound and undeniably changed the course of many battles. Britain, facing the relentless onslaught of the Luftwaffe's bombing campaigns, heavily invested in radar development. Chain Home, a network of ground-based radar stations, formed a crucial part of their defensive strategy along the coast. Its effectiveness in detecting incoming enemy aircraft provided vital information, enabling the timely deployment of fighter squadrons and significantly reducing the effectiveness of German air raids. The subsequent development of airborne radar further enhanced this capacity, allowing fighter pilots to engage enemy aircraft in all weather conditions, a capability previously unimaginable. This technological advantage extended significantly beyond the immediate battlefield.

Naval warfare also saw transformative change with the advent of radar. Ships were no longer reliant on visual observation, particularly in poor weather conditions. Naval radar allowed for the early detection of enemy vessels and submarines, enhancing both offensive and defensive capabilities. Convoys could be routed to avoid enemy patrols, improving the safety of vital supply lines. This capability, to accurately detect submarines, became critical in winning the Battle of the Atlantic, and ultimately, the war. Accurate detection of enemy ships prevented disastrous encounters, saving countless lives and vital resources.

The development of ground-based radar systems, however, extended beyond coastal defenses. These were deployed for a wide range of applications, from artillery fire control to early warning networks. Ground-based systems were capable of tracking artillery shells, using radar to improve the accuracy and effectiveness of counter-battery fire. This dramatically reduced the impact of enemy artillery barrages. The accuracy provided by radar systems allowed for the deployment of counter-battery fire with pinpoint accuracy, minimizing friendly casualties while maximizing effectiveness. Early warning systems ensured the defense forces were alerted to the approach of large-scale enemy advances, allowing for better preparation and deployment of resources.

Furthermore, the race to develop better radar technology spurred significant advancements in related fields, such as electronics and signal processing. The war became a proving ground, driving innovation and pushing the boundaries of what

was technically feasible. This competitive pressure resulted in rapid advancements and created a ripple effect, influencing numerous technological and scientific domains. This intense focus on radar development created a new environment of collaboration between scientists and engineers, resulting in a rapid flow of information and ideas that were pivotal for the entire war effort.
It is important to emphasize the strategic implications of radar. Its deployment wasn't just about improving the effectiveness of individual weapons systems; it fundamentally altered the dynamics of warfare. The ability to see beyond the horizon, to detect unseen threats, granted a significant tactical and strategic advantage, fundamentally shifting the balance of power. This superiority in reconnaissance and situational awareness drastically changed the nature of combat and provided an advantage that could only be countered by the enemy's development of their own radar technology. It wasn't simply a weapon, but a game-changer. The development of radar was a crucial part of the Allied victory, a technology that extended its influence far beyond the battles it directly impacted.
In essence, radar in World War II was not merely a technological advancement; it was a catalyst for change, a force multiplier that fundamentally reshaped the dynamics of modern warfare. From its rudimentary beginnings to its widespread adoption across all branches of the military, radar proved itself invaluable, pushing the boundaries of technology while simultaneously shaping the course of the conflict. Its influence continues to shape our world today, a legacy born from the crucible of war, but now benefiting society in a multitude of ways. The war was won with the help of a vast array of technologies, but the development and application of radar was a cornerstone of Allied success.

2.1 Airborne Surveillance

Airborne surveillance during World War II represented a pivotal moment in radar's history, transforming aerial warfare and fundamentally altering strategic thinking. Early systems, though primitive by today's standards, provided invaluable early warning of incoming enemy aircraft, giving precious time for defensive measures. This capability was revolutionary; previously, the element of surprise held immense power. Now, nations could anticipate attacks, significantly increasing their chances of survival and effective countermeasures. The development of these systems involved a significant challenge: filtering out 'noise' and accurately identifying true threats amidst clutter. Engineers had to grapple with atmospheric interference, ground reflections, and the sheer complexity of interpreting radar returns in a timely manner. The initial successes were often modest, but the potential was undeniable.

These early warning systems were far from perfect; false alarms were frequent, and the accuracy of identifying target type and altitude was limited. Yet, the impact was immediate and profound. Imagine the pilots, suddenly having the ability to see the unseen—to detect enemy aircraft long before they were visually detectable. This ability changed the dynamic of air combat, pushing the need for improved agility and evasive maneuvers. The very nature of aerial warfare began to shift, moving away from relying solely on visual detection. The integration of radar into fighter aircraft provided an additional layer of tactical advantage. Pilots could now engage enemies beyond visual range, gaining a critical strategic edge. The implications were clear: the side with superior airborne radar surveillance would likely

have a decisive advantage on the battlefield.
Furthermore, the development of airborne radar systems fueled intense technological competition. The need for more sensitive receivers, stronger transmitters, and more sophisticated signal processing techniques drove innovation at an unprecedented rate. This competition, born from the necessity of survival, accelerated the pace of technological advancements across various engineering disciplines. The war became a crucible for technological development, pushing the boundaries of what was considered possible and hastening the progress of radar technology far beyond what would have been expected in peacetime. This is evident in the rapid evolution seen in just a few short years, from rudimentary systems to complex networks designed to monitor vast airspace. The pressure of war fostered cooperation and competition—rival nations, though enemies, also spurred each other to develop better technology.
This technological race was not without its limitations. The production of radar systems required significant industrial capacity, a challenge for many nations during the war. The complexity of the technology led to shortages of skilled personnel, creating an additional strain on already stretched resources. This is where the advancement of signal processing algorithms came into the spotlight, aiming to minimize the burden of human interpretation by automating the detection and classification of targets. This meant creating systems capable of automatically identifying friend or foe, a task critical to avoiding friendly fire. The race to develop better algorithms was an important aspect of the technology's evolution, impacting not just military strategies but also the development of computer science itself.
In conclusion, airborne surveillance radar during World War

II was not just a technological advancement; it represented a paradigm shift in military strategy and global geopolitical relations. Its impact extended far beyond the battlefield, fundamentally changing how warfare was conducted and significantly accelerating the development of electronic technology and signal processing. The lessons learned, both in successes and failures, laid the foundation for the continued evolution of radar technology well into the postwar era and its various applications that continue to impact the world today. The technological strides made during this period of intense conflict continue to serve as a potent reminder of both the destructive and constructive power of human ingenuity, especially in a crucible of conflict. The story of airborne radar's war-time deployment is one of strategic advantage gained, scientific innovation fueled by desperate necessity, and a fundamental change in the way humankind understood and engaged in aerial warfare.

2.2 Naval Applications

During World War II, naval applications of radar proved transformative, revolutionizing maritime warfare and significantly altering the balance of power at sea. Before its widespread adoption, naval engagements relied heavily on visual observation and acoustic detection, both severely limited by weather conditions and range. Radar, however, offered an unprecedented advantage: the ability to detect enemy vessels in darkness, fog, or heavy seas. This changed naval strategies fundamentally, enabling earlier detection of threats and more effective responses. The immediate impact was dramatic. Convoy protection, a crucial aspect of the war effort, became significantly safer. The ability to detect approaching submarines and surface ships at greater distances allowed for more timely evasive maneuvers or counterattacks, dramatically reducing losses. The increased situational awareness provided by radar fundamentally altered naval battles. Imagine the sheer advantage of knowing the location, speed, and even potential course of an enemy fleet hours before they were visually detectable. This strategic edge translated into significant military victories.

One particularly impactful development was the integration of radar into fire-control systems. Prior to radar's implementation, aiming naval guns at moving targets was extremely challenging, often resulting in poor accuracy and wasted ammunition. Radar significantly increased the accuracy of targeting and engagement. The immediate result was far greater success rates in targeting and neutralizing enemy vessels, while the long-term effect led to the development of much more advanced fire control systems. Targeting and tracking systems were improved through more precise data. This allowed for more accurate and effective engagement of enemy targets, even during fierce battles.

Furthermore, radar played a critical role in naval navigation, particularly in challenging conditions. Coastal radar systems greatly improved navigation safety by giving ships an accurate view of their position in relation to coastlines, reefs, and other potential hazards. This led to reduced shipwrecks and improved efficiency in convoys. This was even more crucial given the widespread use of mines during the war. Knowing the exact position within a minefield dramatically reduced the likelihood of an accidental detonation.

However, the integration of radar wasn't without its challenges. The technology was still relatively new, and its effective deployment required extensive training and specialized personnel. Moreover, countermeasures were quickly developed by enemy forces to interfere with radar signals, creating the need for evermore sophisticated radar technologies. The constant advancement in technology on both sides resulted in an arms race—a technological battle that pushed the boundaries of radar's capabilities.

The success of naval radar applications during World War II solidified its importance as a critical piece of maritime technology. Beyond the immediate military impact, this experience accelerated the development of more advanced radar systems and set the stage for its widespread adoption in the post-war world. The lessons learned, both in terms of technical innovation and strategic application, were instrumental in shaping the development of future naval capabilities. The development and adoption of radar systems was not only crucial for ensuring the success of naval operations during the war, but also a pivotal advancement that fundamentally altered naval warfare as a whole, establishing a new standard. The improved safety, effectiveness and tactical advantages of naval warfare brought on by radar's integration are a lasting testament to its impact. This evolution, driven by the urgency of wartime necessity, significantly altered the landscape of naval operations and laid the foundation for modern naval technology.

2.3 Ground-Based Systems

Ground-based radar systems, unlike their airborne or naval counterparts, offered a crucial, fixed defensive perimeter during World War II. These systems, often large and imposing, provided early warning of incoming air raids, allowing for precious time to mobilize defenses and prepare civilian populations. Their placement strategically across coastlines and borders created a network of detection, dramatically altering the strategic landscape of the war. The sheer scale of these installations, sometimes encompassing massive antenna arrays and complex processing centers, reflected the technological effort involved in developing and deploying this critical defensive technology. Their impact was profound, impacting not only military strategy but also the civilian experience of wartime.

The development of ground-based radar presented significant engineering challenges. Consider the sheer size of some of these installations, demanding innovative solutions for construction, power supply, and signal processing in an era of limited computing power. The need for reliable operation, even under harsh weather conditions, further amplified the design complexities. Engineers had to overcome significant hurdles in developing robust systems capable of withstanding extreme environmental conditions and maintaining accuracy over long distances. These systems provided the foundation for future advancements in radar technology.

Effective operation depended on a skilled workforce trained in the intricacies of radar operation and maintenance. These personnel weren't simply technicians; they were the interpreters of complex data streams, transforming raw radar signals into actionable intelligence. Their role involved constant monitoring, quick decision-making under pressure, and effective communication with military command. The training programs designed to create this skilled workforce, thus, played a vital role in the effectiveness of ground-based radar. Their actions directly impacted the lives of countless people.

Beyond their military application, the technological groundwork laid by ground-based radar systems profoundly influenced post-war developments. The engineering principles, signal processing techniques, and antenna designs pioneered during this period formed the basis for subsequent advancements in diverse fields. The sheer scale of the investment in these systems created a pool of expertise that flowed into other areas of research and development, generating spin-off technologies impacting civilian life. This legacy extends to numerous technological fields even today.

Interestingly, some of the most significant advancements in ground-based radar technology emerged from the necessity of improving signal processing capabilities. Early systems were limited by the technology available to process large volumes of radar data in real-time. The development of more efficient algorithms and computing power proved crucial in enhancing the accuracy and speed of detecting threats. This drive for improvement resulted in faster processing, more sensitive detection, and a reduction in false alarms, leading to overall system effectiveness and efficiency. These advancements continue to drive innovations today.

Consider the evolution of antenna design. Initial ground-based radar systems employed relatively simple antenna configurations. Over time, however, engineers developed more sophisticated designs to optimize signal transmission and reception. This led to increased range, improved resolution, and the ability to track multiple targets simultaneously. These antenna advancements demonstrated the ongoing process of improvement and technological development within the field of ground-based radar. The implications reached far beyond immediate military applications.

Furthermore, the development of ground-based radar spurred advancements in related fields, such as communications and computing. The need to handle and interpret large volumes of data in real-time fueled research into faster, more efficient computing systems. These advancements had widespread implications, extending far beyond the realm of radar. The synergistic relationship between military necessity and technological innovation is clearly illustrated here. It is a compelling example of how one area of technological development can inadvertently accelerate progress in many others.

In conclusion, the development and deployment of ground-based radar systems during World War II represent a pivotal moment in the history of radar technology. These systems played a crucial role in the war effort, offering a vital defensive capability. Moreover, they spurred significant advancements in related fields, laying the groundwork for numerous post-war applications. The legacy of these systems extends far beyond their military origins, shaping the course of technological development and influencing our world today in many ways. It offers a powerful lesson in the interplay between scientific innovation, military necessity, and technological progress. The impact of this technology continues to this day and remains a significant part of radar history.

3 Post-War Advancements

The immediate post-war era witnessed an explosion of innovation in radar technology, propelled by the vast resources and experience gained during the conflict. Miniaturization, a key driver of this advancement, was paramount. Vacuum tubes, bulky and power-hungry components of wartime radar systems, gradually gave way to smaller, more efficient transistors. This transition not only reduced the size and weight of radar units but also dramatically lowered their power consumption, paving the way for portable and mobile applications previously unimaginable. This shift significantly impacted the development of affordable and widely accessible radar systems.

Consequently, the range of radar applications broadened dramatically beyond the military sphere. Scientists and engineers began exploring its potential in various civilian sectors. Air traffic control, a burgeoning field after the war, quickly embraced radar as an essential tool for managing the increasing volume of air travel. The ability to track aircraft accurately and prevent mid-air collisions became a critical safety measure, profoundly affecting the development of the aviation industry's safety protocols and improving passenger safety.

Improvements in accuracy and reliability were also critical advancements. The need for more precise measurements in diverse fields fueled research into signal processing and data analysis techniques. Engineers developed sophisticated algorithms that enhanced the clarity and accuracy of radar signals, enabling the detection of smaller objects and the differentiation of targets with greater precision. This refinement significantly enhanced the practical utility of radar across various applications, from weather forecasting to navigation systems.

A notable example of post-war civilian application was the development of weather radar. Doppler radar technology, which measures the velocity of moving targets, revolutionized meteorology. For the first time, meteorologists could track the movement of storms with unprecedented accuracy, predicting their paths and intensity with a higher degree of reliability. This improved forecasting ability saved countless lives and minimized damage from severe weather events, representing a significant societal contribution from radar's evolution.

Furthermore, the post-war period saw increased collaboration between government agencies, universities, and private companies in radar research and development. This collaborative environment accelerated technological progress, as experts from diverse backgrounds shared their knowledge and expertise. Government funding fueled many of these advancements, recognizing the strategic importance of radar technology for both national defense and economic growth. The post-war era was therefore not just about technological advancements, but also about a new era of collaborative research.

Beyond air traffic control and meteorology, radar found uses in navigation and surveying. Marine radar systems became standard equipment on ships, improving navigation safety and efficiency. Similarly, ground-based radar systems aided in surveying and mapping, providing accurate measurements of terrain and infrastructure, which played a vital role in construction projects and urban planning. This widespread adoption highlights the adaptability and versatility of radar technology, demonstrating its ability to solve problems in diverse sectors.

The miniaturization of radar components also enabled the development of smaller, more affordable radar systems for scientific research. This allowed scientists to study various phenomena previously inaccessible, with applications ranging from bird migration patterns to atmospheric research. The compactness and lower power consumption also made radar more portable, facilitating research in remote locations.

Finally, the technological leaps made in the post-war years laid the foundation for the continued evolution of radar technology in subsequent decades. The innovations in signal processing, miniaturization, and materials science unlocked new possibilities, paving the way for more advanced and sophisticated radar systems with far-reaching applications. The post-war period truly marked the transition of radar from a niche military technology to a ubiquitous tool transforming various aspects of society.

3.1 Technological Leap

The post-war era witnessed an explosion of innovation in radar technology, moving far beyond its wartime origins. Miniaturization, driven by the development of smaller, more efficient components like transistors and integrated circuits, was key. This allowed for the creation of compact, portable radar systems, expanding their potential applications significantly. No longer confined to large, power-hungry installations, radar became increasingly accessible for civilian use. Think of it: the bulky, room-sized radar sets of the war years gave way to devices small enough to be integrated into aircraft, ships, and even automobiles. This dramatic shift in size and power consumption opened up entirely new avenues for technological development.

The increased affordability and practicality of smaller radar units fueled a rapid expansion into diverse fields. Suddenly, radar wasn't just for military applications; it found its place in air traffic control, guiding planes safely through increasingly busy skies. It enhanced maritime navigation, providing sailors with more accurate positioning and obstacle avoidance. It even began appearing in everyday technologies, though in far more basic and less sophisticated forms, improving the safety and reliability of various systems. This widespread adoption marked a pivotal moment, transforming radar from a specialized military tool into a ubiquitous technology influencing countless aspects of modern life.

The accuracy of radar systems also saw tremendous improvement during this period. Advances in signal processing, the ability to filter out noise and isolate weaker signals, and the development of more sensitive receivers led to the ability to detect targets at greater distances and with finer detail. This enhanced precision enabled new applications, particularly in areas requiring high-resolution imaging. For example, improvements in resolution allowed for clearer pictures of the terrain, leading to improvements in mapping. This, in turn, led to better infrastructure planning and development across the globe. The improvements weren't limited to only resolution; they also enhanced the measurement of speed and direction of movement, paving the way for better traffic management systems.

This period saw the birth of Doppler radar, a revolutionary advancement that could measure the speed of moving objects as well as their position. This capability had immediate and far-reaching consequences, particularly in meteorology. Doppler radar became indispensable for weather forecasting, allowing meteorologists to monitor storm systems with unprecedented accuracy, predicting their movement and intensity, and providing early warnings of severe weather events. This enhanced forecasting capability has undoubtedly saved countless lives and minimized damage from natural disasters worldwide, allowing for better preparedness and effective emergency response strategies. This technology also transformed the understanding of weather systems, leading to a deeper understanding of atmospheric phenomena and better climate modeling.

Moreover, the development of phased array antennas greatly increased the versatility and flexibility of radar systems. These antennas allowed for electronic beam steering, meaning the radar beam could be rapidly scanned across a wide area without physically moving the antenna itself. This capability was crucial in many applications, such as air traffic control, where a single radar system could effectively monitor the movement of numerous aircraft simultaneously. It allowed for more efficient monitoring of a large area, leading to a more efficient use of resources. It was also critical in applications requiring rapid scanning, providing more timely data about the surroundings. This advancement further accelerated the adoption of radar in various sectors and made possible applications that were previously impractical. The technological leap brought about by phased array antennas significantly transformed the capabilities of radar.

Beyond improvements in existing systems, this period also brought about the development of entirely new radar types, each designed to address specific challenges and needs. These new radar systems utilized different frequencies, waveforms, and signal processing techniques to optimize their performance for particular applications. For example, synthetic aperture radar (SAR) emerged as a powerful tool for creating high-resolution images of the Earth's surface, regardless of weather conditions. This opened up new possibilities in various fields, including mapping, surveillance, and environmental monitoring. The development and application of Synthetic Aperture Radar (SAR) fundamentally altered the way we gather information about the Earth's surface, providing high-resolution images even in poor weather conditions, a feat that was once impossible.

The post-war era truly cemented radar's place as a cornerstone of modern technology. It wasn't merely an incremental improvement; it was a fundamental transformation, driven by miniaturization, increased accuracy, and the development of entirely new radar types. The impact extended far beyond the military realm, touching upon almost every facet of daily life. It serves as a powerful example of how a technology born out of necessity can become a ubiquitous tool that improves safety, efficiency, and understanding across a vast array of fields. The legacy of this "technological leap" continues to shape the world around us, influencing scientific exploration and technological advancements in ways that were unimaginable at the time.

3.2 Miniaturization

The post-war era witnessed a relentless pursuit of smaller, more efficient radar systems. Miniaturization wasn't merely a quest for compactness; it unlocked unprecedented possibilities. Smaller radars meant lighter, more portable devices deployable in diverse settings previously inaccessible to their bulky predecessors. This shift fundamentally altered radar's applications, expanding its reach far beyond military contexts.

Imagine the challenges faced by early engineers. Reducing the size of vacuum tubes, the heart of early radar transmitters and receivers, without compromising power or sensitivity was a monumental task. It required innovative approaches to circuit design, material science, and manufacturing processes. The development of transistors in the 1950s provided a breakthrough, paving the way for significantly smaller and more energy-efficient components. This technological leap forward enabled the creation of handheld radar devices – a concept unimaginable just a decade earlier.

Miniaturization also spurred advancements in antenna design. Traditional radar antennas were large, cumbersome structures, often requiring substantial infrastructure for support and operation. The demand for smaller radars pushed engineers to explore new antenna technologies, such as microstrip antennas and phased arrays. These innovative designs allowed for the integration of complex radar systems into compact packages, suitable for use in aircraft, spacecraft, and even hand-held devices.

The impact on cost-effectiveness was profound. Smaller components and simpler designs translated to lower manufacturing costs. This made radar technology accessible to a wider range of applications and users, fostering innovation in both the public and private sectors. The affordability of smaller radar units significantly accelerated their integration into various industries, creating a positive feedback loop where reduced costs stimulated further innovation and miniaturization.

The consequences extended beyond mere economics. Smaller, lighter radar systems increased portability and mobility. Researchers could deploy these systems in remote locations, previously inaccessible to bulky equipment. This allowed for more extensive data collection, leading to new scientific discoveries and insights. For example, smaller, more mobile radar units enabled geologists to map subterranean structures with unprecedented detail, aiding in resource exploration and hazard mitigation. Similarly, ecologists used compact radar systems to monitor wildlife populations and study migration patterns in their natural habitats.

Furthermore, miniaturization opened doors for integration into other technologies. Smaller radar units became integral components in various systems, enhancing their capabilities and functionality. This integration, in turn, resulted in the development of entirely new applications for radar, expanding its influence across multiple disciplines. Think about the navigational systems in modern vehicles; many now incorporate compact radar units for adaptive cruise control and collision avoidance systems. These features were once the realm of science fiction, yet they are now commonplace thanks to the relentless miniaturization of radar technology.

The journey toward miniaturization was not without its challenges. Reducing the physical size of radar systems often meant compromises in performance, particularly in terms of range and

resolution. Engineers had to develop innovative techniques to mitigate these limitations, such as employing advanced signal processing algorithms and improving antenna efficiency. This continuous push to improve performance while maintaining a small footprint fueled a cycle of innovation that continues to shape radar technology today.
The miniaturization of radar is a testament to human ingenuity and the power of persistent innovation. It's not just about making things smaller; it's about unlocking the potential of technology by expanding its accessibility and versatility. From handheld devices to sophisticated integrated systems, the miniaturization of radar has dramatically reshaped its role in scientific discovery and technological advancement, enabling a myriad of applications that continue to transform our world. This ongoing progress is essential for future advancements and paves the way for further exploration of radar capabilities. The reduced size also implies a reduced energy consumption, a significant aspect for sustainable development.

3.3 New Applications

The post-war era witnessed a surge in radar's capabilities, moving beyond military applications and transforming various sectors. Miniaturization, a crucial advancement, enabled the creation of smaller, more portable radar systems, leading to a significant expansion of its use. This shift allowed for integration into diverse applications previously impossible with the large, power-hungry systems of the wartime years. Imagine the leap from bulky, stationary installations to handheld devices capable of performing complex radar functions. This marked a turning

point, paving the way for new markets and research avenues.
Development of sophisticated signal processing techniques drastically improved the resolution and accuracy of radar data. Advanced algorithms allowed for the filtering of noise and the extraction of subtle details from radar returns, leading to a clearer understanding of the information gathered. This enhanced precision opened doors to new fields where high accuracy was paramount, furthering its capabilities far beyond initial expectations. Consider the increased accuracy in measuring the speed and direction of moving objects; this is a testament to the significant strides made in signal processing.

One significant application lies in air traffic control. Modern air traffic management systems heavily rely on radar technology to monitor aircraft movements, ensuring safe and efficient air travel. Radar systems track aircraft position, altitude, and speed, providing air traffic controllers with real-time data vital for managing air traffic flow, preventing collisions, and guiding aircraft safely to their destinations. The complexity of modern air travel wouldn't be possible without the continuous improvement of radar systems.

Beyond air traffic control, maritime navigation greatly benefited from radar's capabilities. Ships use radar systems for collision avoidance, navigation in low-visibility conditions, and mapping of coastal areas. The ability to detect other vessels, even in fog or heavy rain, significantly enhances maritime safety, reducing the risk of accidents and improving efficiency in shipping routes. This application showcases how a technology initially designed for military use has revolutionized safe and reliable transportation.

In the field of industrial process control, radar found a new niche.

Radar sensors are used to monitor levels of liquids and solids in various industrial processes, enabling automated control systems to maintain optimal operating conditions. This application demonstrates radar's versatility, extending its utility beyond just tracking objects; it's now a crucial tool for maintaining consistent operations within various manufacturing processes. These precision measurements are fundamental for a smooth and efficient production line.

Moreover, the development of ground-penetrating radar (GPR) opened entirely new possibilities in subsurface exploration. GPR uses radar pulses to image underground structures, providing information about buried objects, geological formations, and underground utilities. Archeologists, geologists, and civil engineers are among those who benefit from this technology, using it to uncover hidden layers of history, study subsurface formations, and plan construction projects more accurately. This non-destructive technique has revolutionized fields previously limited by the invasive nature of traditional exploration methods.

The application of radar in environmental monitoring is another example of its versatility. Radar systems are used to monitor deforestation, track wildlife populations, and study the effects of climate change. This demonstrates the technology's growing impact in conservation efforts and environmental research. Analyzing radar data allows for a detailed understanding of ecological processes and allows researchers to make informed decisions on environmental management strategies.

Furthermore, specialized radar systems are employed in security applications. These systems detect intrusions, monitor perimeters, and even identify individuals, enhancing security protocols for buildings, infrastructure, and border control. This reflects

radar's contribution to overall public safety. The ongoing advancements in radar technology lead to systems that can provide quicker and more reliable security alerts.
Lastly, the development of synthetic aperture radar (SAR) provided a revolutionary advancement. SAR uses signal processing techniques to create high-resolution images from radar data, even from moving platforms like satellites or aircraft. This technology found wide applications in remote sensing, earth observation, and mapping, providing detailed images of Earth's surface regardless of weather conditions. This capability revolutionized the way we understand and monitor our planet's landscapes. The advancements in SAR provide continuous information about changes in land use, environmental impact, and geological features.

3.4 Civil Applications

The post-war era witnessed a remarkable shift in radar technology's trajectory. No longer solely a tool of military might, its capabilities began to blossom in the civilian sphere, revolutionizing numerous sectors. From enhancing public safety to streamlining industrial processes, radar's versatility proved invaluable.
Air traffic control underwent a profound transformation. The ability to track aircraft precisely, irrespective of weather conditions, became a cornerstone of safe and efficient air travel. Radar systems provided controllers with real-time data, enabling them to guide planes safely through congested airspace and prevent collisions. This marked a significant leap in air travel safety, leading to a considerable reduction in accidents. The devel-

opment of secondary surveillance radar, which transmits and receives signals from transponders on aircraft, significantly improved the precision and reliability of air traffic management.
Navigational systems also benefited immensely. Ships and aircraft utilized radar for navigation in challenging environments, such as fog or storms. Precise location tracking and obstacle avoidance capabilities significantly enhanced safety and efficiency across various transportation sectors. Marine radar systems aided in collision avoidance, while ground-based radar systems guided ships through narrow waterways and harbors. Indeed, the development of advanced radar systems and global navigation satellite systems (GNSS) significantly improved the safety and efficiency of maritime transport.
Beyond transportation, radar technology found applications in diverse fields. Industrial automation saw the integration of radar sensors for monitoring and controlling manufacturing processes. These sensors provided real-time information on the position and movement of objects, enabling precise automation and improved quality control. For example, automated guided vehicles (AGVs) in warehouses and factories relied on radar sensors for navigation and obstacle avoidance. This improved the efficiency of material handling, warehouse management, and other manufacturing processes.
Environmental monitoring became another area of significant impact. Radar systems aided in tracking pollution levels, monitoring the movement of pollutants, and identifying sources of environmental contamination. Coastal erosion monitoring also benefitted, utilizing radar to assess changes in shoreline over time, helping inform coastal management strategies. The development of sophisticated radar systems allowed scientists to

monitor large areas efficiently, providing data that improved environmental protection efforts.

The advancement of speed detection systems transformed law enforcement. Radar speed guns, small, portable devices, improved road safety by providing a fast and accurate method of detecting speeding vehicles. This technology had a considerable impact on reducing traffic accidents and enforcing speed limits. The enhanced accuracy and reliability of radar speed guns also minimized the possibility of inaccurate speed readings.

Furthermore, geological surveys and resource exploration benefited. Ground-penetrating radar (GPR) provided a non-destructive method for subsurface exploration. This technology was used to locate underground utilities, detect geological formations, and explore for mineral resources. The use of GPR significantly reduced the need for extensive and costly excavation, making it a valuable tool in various exploration applications. Its non-invasive nature minimized environmental disruption compared to traditional exploration methods.

In conclusion, the post-war expansion of radar technology into civil applications profoundly impacted various sectors. From air traffic control and navigation to industrial automation and environmental monitoring, radar's versatility transformed numerous industries, enhancing safety, efficiency, and understanding across a wide spectrum of human endeavors. Its non-invasive and precise nature continues to make it a vital tool in countless areas, shaping our world in ways often unseen yet profoundly impactful. The continuing development and refinement of radar systems promises even further integration into our daily lives, promising new solutions to challenges we face in the years ahead.

3.5 Increased Accuracy

The post-war era witnessed a relentless pursuit of enhanced radar accuracy, driven by both military and civilian demands. Improved signal processing techniques, such as matched filtering and pulse compression, dramatically increased the ability to discern faint echoes from background noise, leading to a significant leap in target detection range and precision. This refinement was paramount for applications ranging from air traffic control, where precise location of aircraft is crucial for safe navigation, to weather forecasting, where the accurate measurement of precipitation intensity dictates the effectiveness of storm warnings. Miniaturization played a pivotal role in this accuracy revolution. Smaller, more powerful components allowed for the development of compact, portable radar systems, which were readily deployable in diverse environments. This portability, coupled with improved signal processing, enabled the use of radar in previously inaccessible locations, significantly expanding the scope of its applications and improving the accuracy of data collected in challenging terrains or weather conditions. Consider, for example, the impact on geological surveys, where the ability to deploy compact radar systems in remote areas significantly enhanced the precision of subsurface mapping.

The development of advanced antenna designs further contributed to increased accuracy. Phased array antennas, capable of electronically steering the radar beam, provided improved target resolution and tracking capabilities, minimizing the effects of clutter and interference. This technology revolutionized applications such as air defense systems, allowing for the simultaneous tracking of numerous targets with unprecedented accuracy, and drastically improving the effectiveness of missile guidance systems. The advancements in antenna technology weren't merely incremental; they represented a qualitative shift in radar's capabilities, opening doors to new applications previously deemed impossible.

Beyond hardware advancements, sophisticated software algorithms played a crucial role. Digital signal processing techniques, capable of handling vast amounts of data, became increasingly powerful, allowing for the implementation of sophisticated noise reduction and target discrimination algorithms. This led to a dramatic improvement in the signal-to-noise ratio, resulting in a significant boost in the accuracy of target detection and parameter estimation. One can easily imagine the impact on autonomous vehicles, where accurate object detection is critical for safe navigation, showcasing how software enhancements intertwine with hardware progress to achieve greater accuracy.

Frequency agility, the ability to rapidly switch between different operating frequencies, emerged as a powerful technique to circumvent interference and enhance target discrimination. This capability was particularly valuable in environments with significant clutter, such as urban areas or dense forests. By utilizing multiple frequencies, radar systems could effectively filter out unwanted signals, isolating the target echoes and improving the reliability of measurements. The implementation of frequency agility significantly expanded radar's versatility and accuracy, opening new opportunities in various sectors, from military surveillance to environmental monitoring.

Furthermore, the development of advanced data fusion techniques allowed for the integration of information from multiple radar systems, creating a more comprehensive and accurate picture of the monitored environment. By combining data from different sources, it became possible to overcome limitations of individual sensors and improve the overall reliability and precision of measurements. This synergy among different radar systems and data sources allowed for the creation of more robust and accurate representations of complex environments, leading to improved decision-making in various applications. This development was crucial for improving not just accuracy but also reliability.

The pursuit of higher accuracy continues to drive innovation in radar technology. New materials, improved manufacturing techniques, and the ongoing development of advanced signal processing algorithms promise further improvements in the future. The increased accuracy achieved in radar systems has had a profound impact on various scientific and technological fields, fundamentally transforming our understanding of the world around us and driving advancements in several industries. This continual evolution guarantees that radar will remain a vital tool for scientific discovery and technological progress in the years to come. The future of radar accuracy holds immense potential, shaping the landscape of future technologies and our understanding of various complex systems.

4 Weather Monitoring

Doppler radar, a marvel of modern meteorology, utilizes the Doppler effect – the change in frequency of a wave in relation to an observer who is moving relative to the source of the wave – to measure the velocity of objects, such as raindrops or snowflakes, within a weather system. This allows meteorologists not only to locate storms but also to determine the speed and direction of their movement, providing crucial information for accurate forecasting and timely warnings. Understanding this technology is key to comprehending the sophisticated weather models we rely on daily.

The fundamental principle behind Doppler radar's effectiveness lies in its ability to transmit a microwave signal and then analyze the reflected signal. The shift in frequency of the returned signal directly corresponds to the radial velocity of the target – that is, the speed at which the target is moving towards or away from the radar. This information, coupled with other meteorological data, allows for the creation of detailed weather maps showing not only the location of precipitation but also its intensity and movement patterns. It's a powerful tool, revolutionizing how we anticipate and prepare for severe weather.

Furthermore, the data provided by Doppler radar is instrumental in the development of sophisticated weather prediction models. These models incorporate vast amounts of data, including information from Doppler radar, satellite imagery, surface observations, and numerical weather prediction techniques, to create a comprehensive picture of current and future weather conditions. This allows for improved forecasts, leading to more accurate predictions of severe weather events like tornadoes, hurricanes, and blizzards, ultimately saving lives and protecting property. The sheer volume of data processed and interpreted is astounding and showcases the remarkable capabilities of modern computing.

Beyond simply tracking storms, Doppler radar plays a significant role in understanding the internal structure and dynamics of weather systems. By analyzing the velocity patterns within a storm, meteorologists can identify areas of strong rotation, a key indicator of the potential for tornadoes. This detailed analysis allows for more precise warnings, providing ample time for individuals and communities to take protective measures, minimizing the impact of these potentially devastating events. The precision offered is a testament to the advancements in radar technology.

The applications of Doppler radar extend far beyond the immediate forecasting of severe weather. Its ability to precisely measure precipitation amounts is invaluable for hydrological modeling, helping us understand and predict flooding. This data allows for better water resource management, enabling authorities to make informed decisions about dam operations, flood control measures, and agricultural irrigation. The wide-ranging benefits extend beyond immediate weather predictions.

Moreover, the continuous improvement of Doppler radar technology has led to more accurate and timely warnings. Advancements in signal processing, antenna design, and data analysis techniques have significantly enhanced the accuracy and resolution of radar images. This allows meteorologists to detect smaller and more localized weather events, providing more precise and location-specific warnings. The ongoing development and refinement directly reflect the commitment to improving public safety.

Finally, the impact of Doppler radar on public safety cannot be overstated. The ability to accurately predict and track severe weather events allows for effective dissemination of warnings to the public, enabling timely evacuation, shelter-in-place decisions, and other precautionary measures. This has undoubtedly saved countless lives and mitigated the destructive effects of severe weather around the globe. Its role in protecting communities is undeniably significant. The improved forecasting, coupled with advanced warning systems, highlights the profound influence of Doppler radar on modern society. The technology's evolution continues, promising even greater accuracy and more timely warnings in the future. Its advancements will continue to benefit communities and individuals facing the unpredictable forces of nature.

4.1 Doppler Radar

Doppler radar represents a significant advancement in radar technology, moving beyond simple detection to reveal the velocity of targets. Unlike conventional radar which measures only the range, Doppler radar utilizes the Doppler effect, a change in frequency caused by relative motion between the radar and the target. This allows for the precise measurement of not just where something is, but also how fast it's moving, towards or away from the radar. This is crucial for numerous applications, particularly in meteorology.

The core principle lies in the analysis of frequency shifts in the reflected radio waves. When a target moves towards the radar, the reflected waves are compressed, resulting in a higher frequency. Conversely, movement away from the radar stretches the waves, causing a lower frequency. By meticulously measuring this frequency shift, the radial velocity—the speed along the line of sight—can be calculated. This seemingly small detail opens up a world of possibilities.

Consider the implications for weather forecasting. A standard radar might show a storm's location, but Doppler radar reveals the internal wind speeds within the storm. This is vital for predicting the storm's intensity and potential for severe weather events like tornadoes or flash floods. The ability to pinpoint areas of strong rotational winds, indicative of tornadic activity, is a game-changer in severe weather warning systems, allowing for timely alerts and potentially saving lives. Furthermore, Doppler radar's depiction of wind fields assists in understanding the storm's movement and predicting its path.

Beyond meteorology, Doppler radar finds applications in diverse fields. In traffic monitoring, it measures the speed of vehicles, contributing to improved traffic management and accident prevention. The technology is used in law enforcement, for speed detection, and in air traffic control to track aircraft speed and direction. Furthermore, Doppler radar aids in the study of ocean currents and waves, providing valuable data for oceanographers and marine researchers. The ability to assess the speed of water movement contributes to understanding coastal erosion, marine ecosystems, and the effects of climate change on oceans. Doppler radar's impact extends even to the measurement of wind speeds in wind turbines for optimizing energy production, highlighting its role beyond immediate safety concerns.

The implementation of Doppler radar involves sophisticated signal processing techniques. The received signals are complex mixtures of reflected waves from numerous targets. Advanced algorithms are employed to isolate and process the Doppler shifts from individual targets, filtering out noise and clutter. This often involves techniques like Fourier transforms and pulse-pair processing to isolate Doppler frequencies embedded within the complex radar return. The precision required underscores the importance of signal processing expertise in Doppler radar systems. Data interpretation also necessitates specialized knowledge to translate raw data into meaningful visualizations and actionable insights.

The technology isn't without limitations. Ground clutter, reflections from stationary objects, can obscure moving targets, particularly at close range. Atmospheric conditions such as heavy rainfall can also attenuate the radar signals, reducing accuracy and range. Furthermore, the radial velocity information provides only the component of velocity along the line of sight; the actual speed and direction of the target may need further calculations. Therefore, a deep understanding of the technology's capabilities and limitations is essential for accurate data interpretation and making informed decisions. The challenges inherent in Doppler radar technology continue to drive innovation in signal processing, data analysis, and system design. Developing more robust and reliable systems remains a crucial area of re-

search.
Despite these challenges, Doppler radar has revolutionized our understanding of moving targets. Its impact on our preparedness for severe weather, the safety of transportation systems, and environmental research makes it a cornerstone of modern radar technology. Continued advancements and refinements to its design and signal processing promise to unlock even greater capabilities in the years to come. The pursuit of improved accuracy, range, and reliability underlines its enduring significance in diverse fields that rely on accurate information about moving targets for informed decision-making and operational efficiency. Doppler radar's impact is a testament to the continuous evolution of radar technology and its pivotal role in shaping our understanding of the world around us. The technology's enduring importance is evident in its ongoing contribution to various scientific and technological fields, promising further advancements and applications in the future.

4.2 Storm Tracking

Understanding weather patterns is crucial, and Doppler radar has revolutionized this field. Its ability to detect not only the presence of precipitation but also its velocity is transformative. By analyzing the Doppler shift—the change in frequency of the radar signal caused by the movement of the target—meteorologists can determine the speed and direction of winds within a storm, providing insights into its intensity and potential path. This allows for more accurate and timely warnings, saving lives and minimizing property damage. For example, a rapidly rotating storm cell identified by its high velocity winds, might indicate the presence of a tornado, necessitating immediate action.

The core of storm tracking lies in interpreting radar images. These aren't simple pictures; they are complex representations of data points, each reflecting the strength and movement of precipitation. Learning to read these images requires practice and understanding of the color scales representing different rainfall rates and wind speeds. It's like learning a new language—one that speaks volumes about the atmosphere. A seasoned meteorologist can decipher subtle shifts in color and texture to anticipate changes in the storm's behavior, often predicting its future trajectory with surprising accuracy. Beginners should start with understanding basic color schemes; they indicate the intensity and rate of precipitation.

Beyond simply identifying rain, Doppler radar gives us a three-dimensional perspective on storms. This allows for the construction of detailed models showing the internal structure of the system, showing things like the tilt of the storm, the location of heavy precipitation, and the presence of rotation. Understanding these features is critical for predicting a storm's behavior and issuing the right warnings. For instance, a tilted storm can produce unexpectedly heavy rainfall in a specific area, while a rotating storm cell might signal the potential for a tornado or severe hail. Therefore, this 3D visualization capability is a powerful tool in providing comprehensive situational awareness.

Data analysis is fundamental to effective storm tracking. Raw radar data is voluminous and complex, requiring sophisticated algorithms and software for processing and interpretation. Meteorologists use various software packages to display, analyze, and model this data. These tools allow for the creation of animated loops showing the storm's evolution over time, giving a dynamic view of its development and movement. This is crucial for forecasting because it highlights trends, allowing for a better understanding of what a storm might do.

However, radar technology isn't without limitations. Ground clutter—reflections from stationary objects like buildings and trees—can obscure the radar signal, making it difficult to interpret weather patterns near these areas. This is especially prevalent in mountainous regions or densely populated areas, sometimes obscuring smaller weather systems. Furthermore, attenuation—the weakening of the radar signal as it passes through heavy precipitation—can limit the radar's effective range, causing information loss in the storm's core during heavy downpours. Therefore, we must carefully consider these limitations in our analysis.

Despite these limitations, improvements in radar technology continue. The development of dual-polarization radar provides a more complete picture of precipitation, differentiating between rain, snow, and hail. This allows for better estimations of precipitation type and intensity, leading to more accurate forecasts and better preparation for different types of weather hazards. Advances in data processing and modeling capabilities further enhance our ability to analyze complex storm systems, providing more accurate and timely warnings, and therefore better disaster management preparedness.

To gain a more comprehensive understanding of storm tracking, hands-on experience is invaluable. Many universities offer courses and workshops focused on meteorological data interpretation, and some even provide access to real-time radar data feeds. Participating in these sessions provides invaluable practical experience, allowing beginners to work directly with radar images and interpret various features. Working with a mentor – a seasoned meteorologist who is experienced in data interpretation and radar technology – is a considerable benefit; they can provide guidance and feedback, significantly aiding in the learning process.

Finally, remember that effective storm tracking is not simply about technological prowess. It demands a deep understanding of atmospheric dynamics, weather patterns, and the limitations of radar technology itself. It is a blend of scientific knowledge, technological expertise, and practical experience that forms the foundation of accurate and reliable storm tracking, thus leading to better safety measures and more informed decision-making in the face of severe weather. Continuous learning and improvement are crucial in this constantly evolving field.

5 Astronomy and Space Exploration

Radar, initially conceived for terrestrial applications, unexpectedly found a vital role in the vast expanse of space. Planetary mapping, a cornerstone of our understanding of the solar system, heavily relies on radar's ability to penetrate planetary surfaces. Unlike optical telescopes limited by atmospheric interference and sunlight, radar signals, transmitted from Earth-based antennae or orbiting spacecraft, can pierce through clouds and darkness, revealing subsurface structures and composition. The data received, meticulously analyzed, paints a far more detailed picture of celestial bodies than any other method, unveiling hidden canyons, ice caps, and even evidence of past or present water sources. This detailed information directly influences our understanding of planetary formation and evolution.

From meticulously charting the surface of Venus, a planet perpetually shrouded in thick clouds, to identifying potential subsurface water ice on Mars, radar has repeatedly redefined our knowledge of planetary bodies. The strength and delay of the reflected signal directly correlate to the surface's characteristics: its roughness, density, and the presence of different materials. For instance, the subtle differences in signal reflection from different parts of a moon's surface can point to variations in its composition—evidence of volcanic activity or impact events. This type of data would be impossible to obtain through purely optical methods. Consequently, radar systems allow scientists to create high-resolution 3D maps of planets and moons, adding crucial layers of information to the broader picture.

The tracking of satellites, both in near-Earth orbit and far beyond, is another area where radar plays a critical role. The precision of satellite tracking is paramount; it's essential for accurate navigation, communication, and observation. Radar, with its ability to precisely measure the distance and velocity of objects, is perfectly suited for this task. These measurements are not merely about location; they help predict satellite orbits with incredible accuracy, enabling precise maneuvering and collision avoidance. Imagine the implications if we couldn't accurately track satellites – communication networks would crumble, weather forecasting would falter, and our global navigation systems would fail. The technology's accuracy is crucial for maintaining the integrity of our technology-dependent world.

Moving beyond near-Earth orbits, radar technology reaches into the depths of our solar system. Deep space probes, venturing to distant planets and moons, rely heavily on radar for navigation and scientific data collection. However, the challenges are significantly amplified over vast distances, demanding powerful transmitters and highly sensitive receivers. The faint signals returning from faraway celestial bodies require sophisticated signal processing techniques to extract meaningful data. Despite the hurdles, the rewards are immense. Radar signals sent to and received from distant probes, have provided invaluable data on cometary composition, asteroid sizes and shapes, and the surface properties of icy moons in the outer solar system. This data is essential for planning future missions and refining our understanding of the outer reaches of our solar system.

The limitations of radar should not be ignored. The resolution achievable, particularly at greater distances, is often less than that of high-resolution optical imaging. Atmospheric interference, while less of a problem in space than on Earth, can still impact signal quality. The process also demands significant computational power to process the vast amounts of raw data produced, and the long delay times of signals traveling across interstellar distances present unique challenges. Understanding these constraints is crucial to the continued development of more efficient and powerful systems for deep space exploration. Each technological leap necessitates innovation in signal processing, antenna design, and data analysis. Consequently, our understanding of the cosmos is continually advanced by continuous improvement of radar technology.

In conclusion, the contribution of radar to astronomy and space exploration is undeniable. From detailed planetary mapping to the precise tracking of satellites and the far-reaching probes that explore the outer solar system, radar provides unique insights impossible to achieve through other means. While challenges remain, the continued advancement of radar technology promises even more profound discoveries in the years to come. The combination of powerful transmitters, sophisticated signal processing, and refined data analysis techniques pushes the boundaries of our knowledge about the universe, constantly revealing more and more about the composition and evolution of celestial bodies. The future of space exploration rests, in part, on the shoulders of this technology; continuous innovation in radar technology will undoubtedly pave the way for remarkable discoveries in years to come. The technology continues to evolve, bringing to light previously unseen wonders of our solar system and far beyond.

5.1 Planetary Mapping

Planetary mapping using radar offers a unique perspective on celestial bodies inaccessible by other means. Unlike optical telescopes limited by light, radar penetrates clouds, dust, and darkness, revealing subsurface structures and surface features. This ability is crucial for understanding the geological history and composition of planets and moons. For instance, the topography of Venus, perpetually shrouded in thick clouds, has been extensively mapped using radar, revealing volcanic landscapes and tectonic activity. The resulting high-resolution images provide invaluable insights into the planet's formation and evolution that optical methods simply couldn't achieve.

The process involves transmitting powerful radio waves towards the target, in this case a planet or moon. These waves reflect off the surface, and the time it takes for the echoes to return, along with the strength and frequency shifts of those echoes, provide a wealth of data. By analyzing these signals, scientists can construct detailed three-dimensional maps of the surface, revealing craters, mountains, valleys, and other geological formations. The strength of the reflected signal provides information about the surface material's reflectivity and roughness, helping differentiate between rock, ice, or other substances.

Different radar frequencies penetrate different depths, allowing for subsurface investigations. Low-frequency radar can penetrate deeper into the planetary crust, revealing layers and structures hidden from view. This can unveil information about subsurface water ice, crucial for understanding potential habitability. High-frequency radar, conversely, provides higher resolution surface details, mapping smaller features like boulders or even the textures of various terrains, enhancing our overall understanding of surface composition and processes. The analysis of these data sets demands sophisticated signal processing techniques, often involving algorithms to filter noise and enhance the signal-to-noise ratio, thus maximizing the accuracy and precision of the resulting maps.

Consider the mapping of icy moons like Europa and Enceladus, orbiting Jupiter and Saturn respectively. Radar observations have suggested the presence of subsurface oceans beneath their icy shells, leading to exciting possibilities regarding extraterrestrial life. The ability of radar to penetrate the ice and detect the presence of liquid water, a key ingredient for life as we know it, is a groundbreaking achievement in planetary exploration, highlighting radar's unique contribution to astrobiological research. Future missions will undoubtedly utilize more advanced radar systems with increased power and sensitivity, potentially revealing further evidence of subsurface liquid water and thus advancing our understanding of the potential for life beyond Earth.

The applications extend beyond mere surface mapping. Radar's ability to measure the Doppler shift, or change in frequency of the reflected waves, allows scientists to determine the velocity of surface features, revealing information about processes such as glacial flow, landslides, or even the rotation rates of celestial bodies. This dynamic information further enhances our comprehension of planetary processes and evolution, moving beyond static maps to include the time-dependent aspects of surface activity. Data acquisition for this type of analysis frequently requires carefully planned observational strategies, considering factors such as the distance to the target, the target's rotation, and the characteristics of the radar system itself.

One example of successful planetary mapping through radar is the Cassini mission to Saturn. Its radar instrument played a pivotal role in mapping Titan, Saturn's largest moon, revealing extensive hydrocarbon lakes and seas near the poles. These findings revolutionized our understanding of Titan's unique environment, showing an active hydrological cycle driven by methane rather than water. The high-resolution radar data from Cassini enabled scientists to identify various landforms, assess their distribution and composition, and consequently propose comprehensive models for Titan's geological and climatological processes. These models are still being tested and refined as new data from subsequent missions becomes available.

Furthermore, the technique continuously evolves. Improvements in radar technology, such as the development of more powerful transmitters, more sensitive receivers, and sophisticated signal processing algorithms, lead to increasingly higher-resolution maps and deeper subsurface penetration. The development of advanced radar systems that can simultaneously transmit multiple frequencies or employ different polarization schemes allows scientists to collect a broader range of information, further enhancing the ability to discriminate between different materials and surface structures, thereby leading to a more complete picture of a celestial body's characteristics and evolutionary history.

The future of planetary radar mapping looks particularly promising. Planned missions to icy moons and other potentially habitable celestial bodies will utilize advanced radar systems to search for evidence of subsurface oceans, identify potential biosignatures, and provide crucial information for future human exploration. This technology is instrumental, not only in furthering our understanding of the solar system's formation and evolution, but also in our continued search for extraterrestrial life, pushing the frontiers of our knowledge and reshaping our perspective on our place in the cosmos. The integration of radar data with data from other instruments, such as spectrometers and cameras, further enhances the value of the radar data. A multi-faceted approach combines multiple lines of evidence, providing a more robust and comprehensive understanding of the target body's properties and processes than any single instrument alone could provide. This collaborative approach significantly increases the scientific return of space missions, making such missions more efficient and impactful.

5.2 Satellite Tracking

Precise satellite tracking relies on a sophisticated interplay of ground stations, orbital mechanics, and advanced signal processing. We aren't just talking about pinpointing a satellite's location; we're talking about tracking its precise trajectory, velocity, and even subtle changes in its orientation. This level of accuracy is crucial for many applications, from maintaining precise navigation systems to ensuring the operational health of scientific satellites.

Consider the challenges involved. Satellites are constantly moving, their orbits influenced by numerous factors, from Earth's gravity's irregularities to solar radiation pressure. Ground stations must account for these variables, making precise calculations based on the satellite's known characteristics and observed signals. This involves complex algorithms and powerful computing capabilities. Data from multiple ground stations is often combined to improve the overall accuracy, a process demanding careful synchronization and error correction.

Tracking signals themselves present another hurdle. The signal strength from a satellite diminishes drastically with distance, often becoming weak and noisy. Sophisticated signal processing techniques are vital here. These techniques filter out unwanted noise, extract the relevant information, and correct for atmospheric interference. The design of the satellite's transponder, the device that sends and receives signals, also heavily influences signal quality and the ease of tracking.

Moreover, the choice of tracking method is not arbitrary. Different techniques cater to varying requirements. Passive tracking, relying on observing the satellite's emitted signals, provides a less precise but often more economical method, suitable for satellites with readily available radio beacons. Active tracking, where a ground station sends a signal to the satellite and measures the time taken for the echo to return, offers a far more precise solution but requires a sophisticated transponder onboard the satellite.

Different tracking networks exist, some consisting of a widely spaced network of ground stations offering global coverage, while others employ more geographically concentrated arrays designed for specific missions or regions. The choice of a network's spatial distribution has a direct effect on the accuracy of tracking. The geometry of the ground stations relative to the satellite plays a crucial role in the overall accuracy of the positional estimates, and optimal placement is crucial.

Furthermore, data analysis is a crucial component, beyond the simple tracking process itself. Sophisticated algorithms are required to process the massive datasets generated by global tracking networks, to identify anomalies, to predict future orbits, and to ensure the continued health and integrity of the tracked satellites. This aspect also involves advanced modeling, often using computer simulations to understand and anticipate orbital perturbations. The output from this process goes far beyond simple coordinates; it feeds into trajectory forecasting, mission planning, and collision avoidance, all fundamental to managing the safety and efficacy of space operations.

Finally, the future of satellite tracking is being shaped by advances in various fields. Improved sensors, faster and more powerful computers, and innovative signal processing techniques promise even more accurate and robust tracking. The integration of different tracking systems, creating a more comprehensive and resilient global network, is also actively pursued, offering improved accuracy and reliability. These ongoing advancements ensure that our ability to monitor and manage satellites will continue to improve, paving the way for safer and more productive use of space.

5.3 Deep Space Probes

Deep space probes rely heavily on radar technology for navigation and data acquisition. These probes, venturing far beyond Earth's orbit, face unique challenges not encountered in closer-range applications. The immense distances involved necessitate incredibly powerful transmitters and extremely sensitive receivers, capable of detecting faint echoes bouncing back from distant celestial bodies. Moreover, the long travel times introduce significant delays in signal transmission, demanding sophisticated signal processing techniques to compensate for these delays and extract meaningful data from the returning signals. Unlike terrestrial or near-Earth applications, deep-space radar systems must contend with interstellar interference. Cosmic radiation, solar wind, and other naturally occurring phenomena can mask or distort the faint signals returning from the probe's target. Advanced filtering and noise-reduction algorithms are crucial for extracting a clean signal, allowing for accurate measurements and data analysis. The design of these systems requires careful consideration of antenna size, power consumption, and overall system robustness to ensure reliable operation during the long journey. The reliability of the equipment itself is of paramount importance, with redundancy built in to handle potential malfunctions or equipment failures during the mission. Data interpretation from deep-space probes presents a considerable challenge. The received signals are often weak and degraded by the vast distances traveled. Advanced signal processing techniques, including advanced coding and error correction protocols, are vital for recovering useful information. Scientists employ sophisticated algorithms to remove noise, correct for distortions, and extract the scientific data encoded within the returned signals. This process, known as signal deconvolution, requires significant computational power and expertise. The information extracted can reveal details about the composition, surface features, and even the internal structure of planetary bodies, asteroids, and comets.

The planning and execution of a deep-space probe mission are long-term endeavors involving intricate coordination between engineers, scientists, and mission operators. The mission's parameters, including the target's trajectory, the probe's trajectory, and the data acquisition plan, must be meticulously planned years in advance. The precise timing and targeting of radar pulses are critical for obtaining the most useful information and require the use of highly accurate astronomical data and sophisticated modeling techniques. It is not uncommon for the data to be processed and analyzed for years after the mission is concluded.

Deep space probes equipped with radar technology have provided invaluable data on the solar system and beyond. Missions to Mercury, Venus, Mars, and other planets, including asteroids and comets, have leveraged radar for surface mapping, atmospheric studies, and other scientific investigations. This radar data has revealed vital information about planetary surfaces, including topography, subsurface structures, and the presence of water ice or other significant compounds. The technology employed continues to improve in terms of sensitivity and resolution, promising even more detailed and comprehensive data in future missions.

To maximize the scientific return of deep space probes, engineers continuously strive to improve the sensitivity and resolution of radar systems. Advances in antenna technology, signal processing algorithms, and power generation are crucial for expanding the reach and capabilities of these probes. The development of more compact, energy-efficient systems will enable future missions to explore more distant destinations and gather even more detailed scientific information. The development of robust and highly sensitive antennas is critical for ensuring the success of these missions. This ongoing development helps to ensure more efficient use of resources and expands the possibilities for future exploration of space.

6 Medical Applications

Ultrasound, a cornerstone of modern medical imaging, might seem unrelated to radar, but the underlying principles share striking similarities. Both technologies utilize the transmission and reception of waves to create images of internal structures. In ultrasound, high-frequency sound waves are employed, while radar uses radio waves, a difference in frequency, not fundamental operating principle. This commonality allows for the adaptation of radar's sophisticated signal processing techniques to enhance medical imaging.

Medical radar applications, while less prevalent than ultrasound or MRI, offer unique advantages. For instance, ultra-wideband (UWB) radar systems, operating at much higher frequencies than traditional radar, show promise in detecting subtle changes in tissue properties. These changes could signal the early stages of cancerous growth, potentially leading to earlier diagnosis and improved treatment outcomes. The non-invasive nature of radar also makes it particularly attractive for monitoring patients over time, without the repeated exposure to ionizing radiation associated with techniques like X-rays or CT scans.

One area of intense research involves the use of radar for detecting and monitoring respiratory disorders. By analyzing the subtle shifts in chest wall movement and breathing patterns, UWB radar systems can provide objective measurements of respiratory rate, depth, and even the presence of obstructions within the airways. This data could be crucial in managing conditions like asthma, chronic obstructive pulmonary disease (COPD), and sleep apnea, enabling timely interventions and personalized treatment strategies. Moreover, continuous, non-intrusive monitoring using radar could revolutionize home healthcare, allowing patients to be remotely monitored for potential respiratory issues.

Beyond respiratory monitoring, researchers are exploring the application of radar in detecting and monitoring cardiac activity. The ability to non-invasively monitor heart rate and rhythm using radar opens exciting possibilities for continuous heart rate monitoring during exercise, sleep, or even daily activities. This could dramatically improve the management of cardiac conditions, enabling early detection of arrhythmias and providing valuable data for personalized treatment strategies. This technology could benefit patients with implanted cardiac devices, eliminating the need for repeated invasive procedures for monitoring purposes.

Another fascinating area is the use of radar for breast cancer detection. While still under development, researchers are investigating the potential of UWB radar to detect subtle changes in breast tissue characteristics indicative of cancerous tumors. The ability of radar to penetrate tissues and detect variations in dielectric properties makes it a promising tool for detecting lesions that might be missed by other imaging methods. This could lead to early detection and improved survival rates, potentially saving thousands of lives every year. Significant advancements in

signal processing and image reconstruction are necessary, however, to make this a reliable and widely used clinical tool.
The challenges in applying radar to medical applications are numerous. The relatively low resolution compared to other imaging modalities remains a significant limitation. The accuracy of radar-based diagnostics is influenced by factors like body composition and movement artifacts, which can interfere with signal interpretation. However, ongoing advancements in signal processing, antenna design, and machine learning techniques are addressing these challenges. Sophisticated algorithms are being developed to filter noise, improve resolution, and automate the interpretation of radar signals, enhancing the reliability and efficiency of radar-based medical diagnostics. These innovations show great promise for overcoming existing limitations and fully unlocking the potential of radar in healthcare. Future advancements hold the key to integrating radar systems seamlessly into clinical workflows.

6.1 Imaging Techniques

Medical radar imaging, unlike traditional X-rays or CT scans, utilizes radio waves instead of ionizing radiation. This non-invasive approach offers a significant advantage, particularly for patients who require frequent scans or have underlying health conditions that make them more susceptible to radiation exposure. The technology operates on the principle of backscattered radio waves, where transmitted signals reflect off internal tissues and organs, providing valuable data on their structure and composition. Signal processing algorithms meticulously analyze these reflected waves, constructing detailed images that aid in

diagnosis.
A key application lies in detecting and characterizing soft tissues, an area where traditional imaging techniques sometimes struggle. For example, radar imaging shows promise in breast cancer detection, allowing for a more precise identification of tumors and assessment of their extent, potentially improving treatment planning and outcomes. Similarly, the technology's sensitivity to minute changes in tissue density makes it a potential tool for monitoring the progression of various diseases, from cardiovascular issues to neurological disorders. However, current resolutions, although improving rapidly, are still lower compared to established medical imaging modalities like MRI or ultrasound.
One exciting development is the use of ultra-wideband (UWB) radar. UWB radar transmits signals over a broader frequency range, achieving significantly higher resolution than traditional radar systems. This improvement is critical in medical imaging, as it enables better differentiation between various tissue types and the detection of smaller abnormalities. Imagine the potential for early detection of tumors, aneurysms, or other life-threatening conditions. This improved resolution is not just a theoretical improvement; ongoing clinical trials are demonstrating the effectiveness of UWB radar in various medical applications. The technology has even proven useful in non-invasive monitoring of vital signs like respiration and heart rate, offering a continuous stream of real-time data, potentially revolutionizing patient monitoring.
Despite its advantages, challenges remain. Penetration depth of radio waves in biological tissues limits the use of radar in imaging deep-seated organs, posing a significant limitation.

Moreover, processing the massive amount of data generated by high-resolution radar systems requires significant computational power, introducing additional complexities. Further development in signal processing algorithms and the use of advanced materials, such as metamaterials with enhanced wave-guiding properties, are crucial to address these limitations. The development of sophisticated algorithms for noise reduction and artifact removal are also key areas of research. These algorithms are vital for ensuring the accuracy and reliability of medical images produced by radar systems.

Moreover, the widespread adoption of medical radar imaging requires extensive clinical validation and regulatory approval. Rigorous testing and validation are necessary to establish the technology's efficacy and safety. Furthermore, the cost-effectiveness of the technology must be demonstrated to ensure its accessibility. A significant investment in research and development is crucial to overcome these challenges and pave the way for wider clinical adoption. The integration of radar imaging with other diagnostic techniques could be particularly beneficial, potentially offering a more comprehensive and accurate assessment of a patient's health status. Such a multi-modal approach would combine the unique strengths of various imaging modalities, minimizing the limitations of each individual technique. Imagine combining radar's sensitivity to soft tissues with ultrasound's high temporal resolution— a truly groundbreaking capability.

Ultimately, the application of radar imaging in medicine is still a rapidly evolving field. The potential benefits are substantial, ranging from early disease detection to improved treatment planning. While significant hurdles remain, ongoing research and development promise to overcome many of the current limitations. Continued advancements in signal processing, materials science, and clinical validation are essential steps in realizing the full potential of radar in revolutionizing medical diagnostics and patient care. The future of medical radar imaging is bright, offering a powerful and non-invasive tool for improving human health. Further refinement of the technology will undoubtedly lead to its widespread adoption and transform the landscape of medical imaging.

6.2 Diagnostic Tools

Ultrasound, a cornerstone of modern medical diagnostics, leverages high-frequency sound waves to create images of internal organs and tissues. Unlike X-rays, ultrasound is non-ionizing, meaning it doesn't use radiation that could damage cells. This makes it a particularly safe and versatile imaging modality, especially for pregnant women and children. The technique works by transmitting sound waves into the body; these waves bounce off different tissues and return to the transducer, a device that both transmits and receives the signals. The time it takes for the waves to return, coupled with their intensity, allows a computer to construct a detailed image of the body's interior. Differences in tissue density, such as those between solid organs and fluids, create contrasts in the resulting image, enabling clinicians to identify abnormalities or pathology.

Microwave radar, while less frequently used in medical imaging than ultrasound, offers a unique set of capabilities. Its ability to penetrate deeper into tissues compared to ultrasound makes it suitable for certain diagnostic tasks, although image resolution is generally lower. Microwave radar technology finds applications in detecting breast cancer, analyzing tissue composition, and monitoring the healing of wounds. Researchers are actively investigating ways to enhance the resolution and image quality of microwave radar systems, striving to bring this powerful tool more into the mainstream of medical diagnostics. Its non-invasive nature, along with its potential for early disease detection, makes this technology extremely promising.

The principle behind radar imaging, whether in ultrasound or microwave applications, is the precise measurement and interpretation of reflected signals. Careful calibration and advanced signal processing techniques are essential for accurate interpretation. This requires sophisticated algorithms and powerful computing resources to transform raw signals into meaningful medical images. The accuracy of the diagnostic information directly depends on the quality of the equipment, the expertise of the operator, and the robustness of the data analysis methods. Furthermore, the understanding of the underlying physics is crucial for selecting the most appropriate radar technology for a particular application.

One significant limitation of radar imaging in medicine, however, is the difficulty in distinguishing between various soft tissues. Often, the contrasts between different tissue types are subtle, leading to ambiguity in interpretation. Consequently, radar imaging is seldom used as a standalone diagnostic tool. Instead, it is often employed in conjunction with other imaging methods, such as X-rays, MRI, or CT scans, to provide a more comprehensive and accurate diagnosis. This multi-modal approach helps clinicians integrate data from multiple sources, reducing uncertainty and improving the overall accuracy of diagnosis.

Beyond imaging, radar technology's role in medical diagnostics extends to other, less visually-oriented applications. For instance, the precise measurement of tissue electrical properties using radar technology allows doctors to monitor the physiological state of tissues, identifying potential problems in areas like wound healing or even detecting early signs of diseases. This method is more of a non-invasive remote sensing of vital signs than a visualization technique, but it is a valuable part of the diagnostic process. It holds promise for the development of new early detection techniques, particularly for conditions that may be difficult to diagnose with other established methods. This is an area of ongoing research, with potentially transformative applications in the future.

The continued development and refinement of radar-based diagnostic tools necessitate a multidisciplinary approach. Collaboration between medical professionals, physicists, engineers, and computer scientists is crucial for advancing this field. Improved signal processing algorithms, advanced sensor technology, and a deeper understanding of the interaction of radar signals with biological tissues are all necessary to overcome current limitations and realize the full potential of radar in medical diagnostics. This calls for innovative interdisciplinary programs, integrating the latest findings from various scientific fields. The future of radar-based diagnostics is promising; further research and developments will undoubtedly lead to even more sophisticated and accurate tools.

7 Future of Radar Technology

The next generation of radar systems promises a quantum leap in capability. We're moving beyond simple detection; we're entering an era of highly precise, sophisticated sensing. Imagine radar capable of identifying not just the presence of an object, but also its composition, its internal structure, even its operational status—all with unprecedented accuracy and speed. This is the reality we are rapidly approaching. Miniaturization, too, continues at a breathtaking pace, allowing for the deployment of radar technology in smaller, more versatile platforms—from tiny drones to implantable medical devices.

This enhanced precision opens up entirely new application areas. Consider autonomous vehicles, for example. The current reliance on lidar and camera systems for object detection is just the beginning. Future vehicles will undoubtedly incorporate advanced radar systems offering superior performance in challenging weather conditions, leading to safer and more efficient autonomous navigation. The integration of sophisticated signal processing techniques will dramatically reduce false positives and greatly improve the overall reliability of self-driving systems. We are talking about a complete paradigm shift in automotive safety and efficiency.

The synergy between radar and artificial intelligence is particularly exciting. AI algorithms can analyze vast quantities of radar data in real-time, identifying patterns and anomalies that would be invisible to human operators. This is particularly beneficial in security and surveillance applications, where AI-powered radar systems can autonomously detect and classify threats, providing early warning of potential dangers. Furthermore, AI can enhance the performance of existing radar systems by dynamically optimizing signal parameters and processing strategies based on real-time conditions, resulting in increased sensitivity and resolution. The potential for improved decision-making and automated responses is enormous.

Beyond the immediate applications, the long-term potential of radar is truly remarkable. We are witnessing breakthroughs in areas such as terahertz radar, which offers the possibility of incredibly high-resolution imaging, enabling the detection of extremely small objects or subtle changes in materials. This technology holds immense promise for applications ranging from medical diagnostics to security screening. Moreover, advances in metamaterials are poised to revolutionize antenna design, leading to smaller, more efficient, and more versatile radar systems capable of operating at a wider range of frequencies.

The development of advanced radar sensors continues to drive innovation across multiple disciplines. However, as radar technology becomes increasingly sophisticated, ethical considerations must remain at the forefront. The potential for misuse, for instance, in surveillance or the collection of personal data, must be addressed. We need to establish clear guidelines and

regulations to ensure the responsible development and deployment of these powerful technologies. This requires proactive collaboration between researchers, policymakers, and the public to guarantee that advancements in radar technology benefit humanity as a whole.
Ultimately, the future of radar technology is characterized by a convergence of technological advancements, from the development of novel sensors and materials to the integration of sophisticated algorithms. These developments are not isolated; rather, they are interconnected, creating a synergistic effect that pushes the boundaries of what's possible. The implications are profound, extending across a vast spectrum of applications, from improving weather forecasting and revolutionizing autonomous vehicles to enhancing medical imaging and advancing our understanding of the cosmos. The trajectory is clear: radar technology continues to evolve at an impressive rate, promising transformative impact on countless aspects of our lives. The challenge ahead lies in harnessing its power responsibly, mitigating potential risks, and ensuring its benefits are shared equitably.

7.1 Advanced Sensors

The development of advanced sensors represents a pivotal advancement in radar technology, pushing the boundaries of what's achievable. Think of miniaturization; we've moved from bulky, power-hungry systems to incredibly compact and efficient devices. This allows for deployment in previously impossible locations, from tiny drones monitoring infrastructure to implantable medical sensors. Consider the shift toward multi-functional sensors: a single unit can now perform multiple tasks simultaneously, gathering data across a range of frequencies and modalities. This integrated approach provides richer, more nuanced information, significantly improving data analysis and interpretation. Increased sensitivity allows detection of smaller, more distant targets, expanding applications from detecting micrometeoroids in space to monitoring subtle changes in the Earth's climate.

These improvements aren't simply incremental; they represent a paradigm shift. The precision achievable with modern sensors is extraordinary. We're talking about resolving targets with sub-millimeter accuracy, a level of detail previously unimaginable. This advancement significantly impacts various fields, from highly precise mapping and navigation to the development of sophisticated medical imaging techniques. The resulting data demands sophisticated algorithms for processing and analysis, opening new doors for data visualization and interpretation, leading to more accurate and efficient systems.

Moreover, new materials and manufacturing processes have played a significant role. Advanced materials, such as metamaterials and novel semiconductors, enable the creation of sensors with unprecedented performance characteristics. These materials allow for improved signal processing, increased bandwidth, and greater resistance to interference, all of which significantly enhances the overall performance and reliability of the system. These advancements have revolutionized manufacturing processes; consider the advent of 3D-printing techniques, enabling the creation of complex sensor geometries previously impossible to achieve using traditional methods. This has resulted in more efficient and cost-effective production, paving the way for widespread adoption and deployment.

The integration of advanced signal processing techniques is equally critical. Sophisticated algorithms enable the extraction of meaningful information from complex radar signals, even in the presence of significant noise or interference. This includes advanced filtering techniques, signal compression, and the use of machine learning algorithms to identify patterns and anomalies. These signal processing advances are essential for effectively utilizing the increased data volume generated by advanced sensors. Such algorithms help to filter out noise, enhancing the signal-to-noise ratio, making the data clearer and more easily interpretable. This allows for more efficient data analysis, which contributes to improved overall accuracy and reliability.

Ultimately, the development of advanced sensors will continue to drive innovation across various sectors. From enhancing weather forecasting accuracy to enabling new medical diagnostic capabilities, the potential applications are vast and far-reaching. The ongoing research and development efforts in this area promise to deliver even more sophisticated sensors with improved performance and new functionalities in the coming years. It's a dynamic field, constantly evolving, constantly pushing the boundaries of what is possible. Consider the future impact – advancements in sensor technology will inevitably revolutionize fields ranging from environmental monitoring to autonomous navigation systems, profoundly affecting our daily lives in ways we can only begin to imagine.

7.2 AI Integration

The integration of artificial intelligence (AI) promises to revolutionize radar technology, pushing its capabilities far beyond what was previously imaginable. We're moving beyond simply detecting objects; AI allows us to interpret the vast amounts of data generated by modern radar systems with unprecedented speed and accuracy. This opens up a wealth of new possibilities. Consider the advancements in signal processing: AI algorithms can sift through noisy data, identifying subtle patterns and anomalies that would be missed by human analysts or traditional methods. This increased sensitivity can lead to earlier detection of threats, more precise weather forecasting, and the ability to discern fine details in astronomical observations. Furthermore, the ability to learn and adapt is crucial. AI systems can continually refine their performance based on new data, becoming increasingly effective over time. They can autonomously adjust parameters to optimize performance in changing environmental conditions, from atmospheric interference to shifting terrain.

AI also offers significant improvements in target classification. Instead of just identifying the presence of an object, AI-powered radar systems can analyze its characteristics—size, shape, speed, even material composition—to classify it with far greater accuracy. This is especially crucial in military applications, where quickly distinguishing friend from foe is paramount. But the benefits extend beyond defense. In weather forecasting, more precise classification of precipitation types—hail, rain, snow—enhances the accuracy of predictions and helps in developing more targeted emergency response strategies. Even in medical imaging, AI could assist in identifying subtle variations in tissue density, aiding in the earlier diagnosis of diseases. The potential applications are virtually boundless.

The challenge lies not just in developing the AI algorithms themselves, but in effectively integrating them into existing radar systems and workflows. This requires careful consideration of data handling, computational power, and the human-machine interface. It's not simply a case of adding an AI component to a traditional system; it necessitates a holistic redesign, often involving specialized hardware and software architectures. Think of the volume of data that modern radar systems generate—a sheer deluge that would overwhelm any human analyst. AI is essential to process this data in real-time, extracting meaningful insights and presenting them to operators in a clear and concise manner. This streamlined workflow translates into faster decision-making and more efficient resource allocation, regardless of the specific application.

One particularly exciting area is the development of autonomous radar systems. These systems, controlled by sophisticated AI, can operate independently, adapting to changing conditions and making decisions without human intervention. Imagine autonomous drones equipped with AI-powered radar, capable of navigating complex environments and performing tasks such as search and rescue operations or environmental monitoring. Similarly, AI-controlled radar systems could manage air traffic autonomously, optimizing flight paths and preventing collisions. However, the development of such systems presents significant engineering challenges, requiring robust AI algorithms and fail-safe mechanisms to ensure reliability and safety. This raises critical ethical questions surrounding accountability and decision-making authority in such systems.

The future of radar hinges on our ability to effectively harness the power of AI. Its potential to enhance accuracy, speed, and autonomy is undeniable. Yet, responsible integration requires careful planning, rigorous testing, and ongoing evaluation to ensure that these powerful tools are used ethically and effectively. As we push the boundaries of radar technology with AI, we must also consider the potential societal impacts and implement safeguards to mitigate any risks. This includes addressing issues of data privacy, algorithmic bias, and the potential for misuse of autonomous radar systems. The responsible development and deployment of AI in radar technology requires a collaborative effort among researchers, engineers, policymakers, and the broader community. Only through careful consideration of these factors can we fully realize the transformative potential of AI-integrated radar systems and ensure that they benefit society as a whole.

7.3 Emerging Trends

Miniaturization continues to be a driving force, pushing the boundaries of what's possible. We're seeing the development of extremely compact radar systems, integrated into everyday devices like smartphones and even wearable technology. Imagine a world where your phone can detect approaching vehicles before you can see them, or a smartwatch that monitors your vital signs with unprecedented precision using micro-radar. This isn't science fiction; it's rapidly becoming a reality, fueled by advancements in microelectronics and antenna design. The implications for personal safety and health monitoring are vast, promising a future where technology seamlessly integrates into our lives for both convenience and enhanced safety.
The fusion of radar with other sensing technologies is creating powerful hybrid systems that provide a more comprehensive pic-

ture of the environment. By combining radar with LiDAR, for example, we can achieve superior object detection and classification, especially in challenging conditions like fog or rain where vision systems often fail. This synergy allows for more robust autonomous driving systems, enhancing safety and reliability. Furthermore, the integration with thermal imaging enables the detection of objects based on their heat signature, expanding the capabilities beyond simple visual or radar-based detection. This multifaceted approach to sensing is crucial for applications such as advanced driver-assistance systems (ADAS), robotic navigation, and environmental monitoring.

Beyond the technological advancements, there's a growing emphasis on energy efficiency. Researchers are exploring novel designs and materials to reduce power consumption in radar systems. This is particularly crucial for battery-powered devices and applications requiring long operational times. Advances in low-power electronics, signal processing techniques, and efficient antenna designs are contributing to this trend. Imagine a long-range weather radar operating for weeks on a single battery charge, vastly expanding its utility in remote areas and difficult-to-access environments. This focus on energy efficiency is not just about cost savings; it also contributes to environmental sustainability, making radar technology more environmentally friendly.

The development of new frequency bands for radar operation is opening up a whole new spectrum of possibilities. Millimeter-wave and terahertz radar offer superior resolution and bandwidth, leading to more accurate and detailed measurements. These higher frequencies allow for more precise detection of smaller objects, improving the effectiveness of applications ranging from medical imaging to airport security. Furthermore, the exploration of extremely high frequency (EHF) radar is unlocking potential for unprecedented capabilities in atmospheric sensing and material characterization, pushing the boundaries of scientific discovery. The exploration of these new bands, however, requires overcoming challenges related to signal attenuation and atmospheric interference.

Another major trend is the increasing use of sophisticated signal processing techniques. Advanced algorithms are enabling better clutter rejection, target identification, and data interpretation. Machine learning and artificial intelligence, while not the focus of this chapter, are playing a growing role in enhancing the performance and efficiency of radar systems. The ability to analyze complex radar data in real time, providing near-instantaneous interpretations, unlocks enhanced capabilities. This advancement isn't just about faster processing speed; it's about extracting meaningful insights from raw data that might have previously been overlooked. This leads to improved decision-making in critical applications such as air traffic control and autonomous navigation.

Finally, cybersecurity is becoming a critical consideration in radar technology. As radar systems become more sophisticated and integrated into critical infrastructure, protecting them from cyberattacks is paramount. The development of robust security protocols and encryption techniques is essential to ensure the reliability and integrity of radar data. This is particularly relevant for applications such as military surveillance, power grid monitoring, and air traffic management, where a compromise could have severe consequences. Ongoing research focuses on securing communication pathways and strengthening the software that governs radar systems' operations, safeguarding the integrity of data and preventing malicious actions. The future of radar rests not just on technological advancements but also on the ability to maintain a secure and reliable operational environment.

7.4 New Frontiers

The inherent limitations of current radar systems, particularly their susceptibility to environmental interference like rain and atmospheric anomalies, represent a significant hurdle. Overcoming this requires a paradigm shift, moving beyond simple signal reflection analysis towards more sophisticated data interpretation methods. Think of it like trying to hear a whisper in a hurricane – current methods struggle, but future developments might involve advanced signal processing techniques, perhaps incorporating machine learning algorithms to filter out the noise and isolate the relevant signals, delivering clearer, more accurate readings even in challenging conditions. We are talking about a transformation similar to the shift from simple photographic plates to digital image processing in astronomy, opening new doors for observation and data analysis. This will inevitably lead to higher resolution and more reliable data, paving the way for advancements across various fields.
Miniaturization has unlocked incredible potential, but truly next-generation radar demands a leap beyond simply shrinking existing components. The focus needs to shift towards novel materials and manufacturing processes that allow for the creation of highly efficient, low-power, and extremely sensitive radar systems. Imagine sensors so small they could be integrated directly into clothing, providing continuous environmental monitoring or

even used for advanced biometric identification. The development of metamaterials, for example, offers potential for creating antennas with unprecedented properties, allowing for smaller footprints and improved performance. This will undoubtedly lead to applications currently deemed science fiction, from advanced personal safety systems to revolutionary environmental monitoring networks.

Beyond miniaturization, the pursuit of enhanced spatial resolution remains a critical area for exploration. Current systems often struggle to distinguish between closely spaced objects, particularly in complex environments. Improved resolution, however, would dramatically enhance the capabilities of radar in diverse applications, from autonomous vehicle navigation to medical imaging. Consider the potential benefits for autonomous vehicles: higher-resolution radar could allow for much safer and more precise navigation, especially in dense urban settings or inclement weather. This would need advancements in signal processing and antenna design, possibly involving techniques like synthetic aperture radar (SAR) with much higher resolution and more sophisticated processing algorithms to extract finer details from the reflected signals. The result would be a significant leap in capability, impacting numerous fields.

The integration of advanced signal processing techniques and innovative antenna design represents an essential step towards unlocking the full potential of radar. This includes exploring new frequency bands, such as terahertz, which offer unique advantages for high-resolution imaging and material identification. Imagine the possibilities: a terahertz radar system could identify hazardous materials at a distance, or provide a non-invasive method of inspecting infrastructure for structural weaknesses.

However, the development of efficient and reliable terahertz sources and detectors poses a significant challenge, requiring innovative solutions in materials science and engineering. The payoff, though, could be transformative.
Finally, the ethical considerations surrounding the increasing capabilities of radar systems must be addressed proactively. As radar technology becomes more powerful and sophisticated, its potential for misuse must be carefully considered. This necessitates the development of robust regulations and guidelines to ensure that radar technology is used responsibly and ethically, preventing its use for malicious purposes. This requires a proactive approach, involving collaboration between researchers, policymakers, and the public to establish ethical frameworks for the responsible development and deployment of this powerful technology. This responsible development is crucial for the future of the field.

7.5 Ethical Considerations

The increasing sophistication of radar technology necessitates a serious examination of its ethical implications. We've seen its transformative power across various sectors, from weather forecasting to medical diagnostics. But this power comes with responsibilities that we, as developers, users, and members of society, must actively address. The potential for misuse, intentional or unintentional, demands careful consideration.
Consider the privacy implications of highly sensitive radar systems capable of detecting minute details. The potential for unwarranted surveillance, whether by governments or private entities, presents a significant ethical challenge. Clear guidelines and

robust regulations are crucial to prevent the abuse of such powerful technology and protect individual privacy rights. Without these safeguards, the potential for harm is substantial. This calls for a multi-faceted approach, involving not just technological solutions but also legal frameworks and public discourse.

Furthermore, the accuracy and reliability of radar data, while impressive, are not infallible. Mistakes in interpretation or malfunctions in the equipment could have severe consequences, particularly in applications impacting public safety, such as air traffic control or weather forecasting. The developers bear a significant ethical burden to ensure the highest standards of quality control and rigorous testing to minimize the risk of errors. Furthermore, transparent communication about the limitations of the technology is essential to prevent misplaced trust and the potential for misinformed decision-making. This includes clearly outlining the margin of error and the potential for false positives or negatives.

The environmental impact of radar technology should also not be overlooked. While radar technology itself might not directly pollute, the energy consumption involved in its operation, particularly with large-scale systems, can be considerable. There's an ethical imperative to explore and implement energy-efficient designs and sustainable practices in the development and deployment of radar systems, minimizing our overall environmental footprint. This involves continuous research and development into more energy-efficient components and power sources, and potentially the exploration of alternative technologies where feasible. This commitment to sustainability is not simply a matter of environmental responsibility, but an ethical obligation to future generations.

Military applications of radar raise complex ethical dilemmas. While radar plays a critical role in defense systems, it also raises questions about the potential for escalation of conflicts and the indiscriminate targeting of civilians. A thorough ethical review process for military radar applications is essential, ensuring that these systems are employed responsibly and in compliance with international humanitarian law. This requires an ongoing dialogue between technological experts, policymakers, and ethicists to develop appropriate safeguards and limitations, preventing the technology's potential for misuse or unintended harm. This includes incorporating ethical considerations throughout the entire lifecycle of these weapons systems, from research and development to deployment and decommissioning.

Beyond these specific concerns, a broader ethical framework is needed to guide the development and use of radar technology. This requires open public discourse, involving scientists, policymakers, and the public at large. It's important to develop a shared understanding of the ethical implications of this technology, establishing a consensus on acceptable uses and limits. Promoting transparency and accountability in the development and use of radar systems will further enhance public trust and ensure its responsible implementation. This ongoing conversation is critical to navigate the ethical landscape and ensure that this powerful technology benefits humanity while mitigating its potential risks. It necessitates a proactive and collaborative approach, ensuring that ethical considerations are central to the continuous development and application of radar technology. The future of radar is not simply about technological advancements, but also about the ethical choices we make today.

www.ingramcontent.com/pod-product-compliance
Lightning Source LLC
LaVergne TN
LVHW022018200926
848481LV00007B/2450